WATCH ANI...

Ducklings

by Colleen Sexton

BLASTOFF! READERS

BELLWETHER MEDIA • MINNEAPOLIS, MN

Note to Librarians, Teachers, and Parents:

Blastoff! Readers are carefully developed by literacy experts and combine standards-based content with developmentally appropriate text.

Level 1 provides the most support through repetition of high-frequency words, light text, predictable sentence patterns, and strong visual support.

Level 2 offers early readers a bit more challenge through varied simple sentences, increased text load, and less repetition of high-frequency words.

Level 3 advances early-fluent readers toward fluency through increased text and concept load, less reliance on visuals, longer sentences, and more literary language.

Level 4 builds reading stamina by providing more text per page, increased use of punctuation, greater variation in sentence patterns, and increasingly challenging vocabulary.

Level 5 encourages children to move from "learning to read" to "reading to learn" by providing even more text, varied writing styles, and less familiar topics.

Whichever book is right for your reader, Blastoff! Readers are the perfect books to build confidence and encourage a love of reading that will last a lifetime!

This edition first published in 2010 by Bellwether Media, Inc.

No part of this publication may be reproduced in whole or in part without written permission of the publisher. For information regarding permission, write to Bellwether Media, Inc., Attention: Permissions Department, 5357 Penn Avenue South, Minneapolis, MN 55419.

Library of Congress Cataloging-in-Publication Data
Sexton, Colleen A., 1967-
  Ducklings / by Colleen Sexton.
     p. cm. — (Blastoff! readers: watch animals grow)
  Summary: "A basic introduction to ducklings. Simple text and full color photographs. Developed by literacy experts for students in kindergarten through third grade"–Provided by publisher.
  Includes bibliographical references and index.
  ISBN-13: 978-1-60014-388-5 (paperback : alk. paper)
  1. Ducklings–Juvenile literature. I. Title.

QL696.A52S49 2008
598.4'1139–dc22
                                            2007040272

Text copyright © 2010 by Bellwether Media, Inc. BLASTOFF! READERS and associated logos are trademarks and/or registered trademarks of Bellwether Media, Inc. Printed in the United States of America.

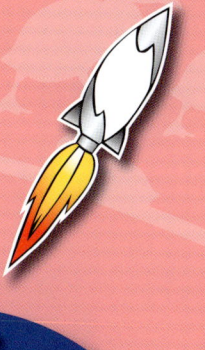

Contents

Newborn Ducklings	4
Parts of a Duckling	8
What Ducklings Eat	16
Ducklings Grow Up	20
Glossary	22
To Learn More	23
Index	24

Crack!
Ducklings **hatch** from eggs. Ducklings are baby ducks.

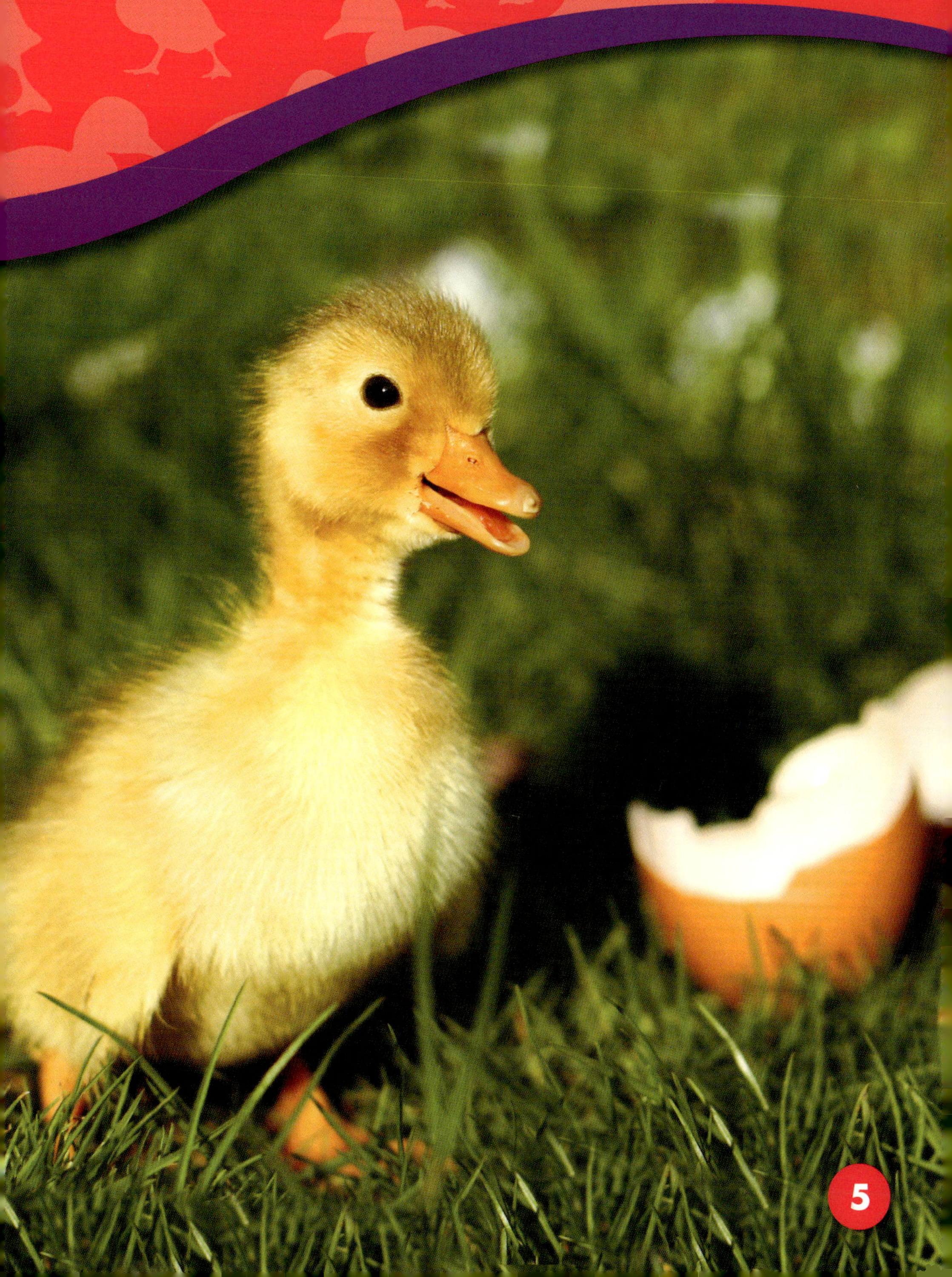

Ducklings stay close to their mother. Their mother keeps them safe.

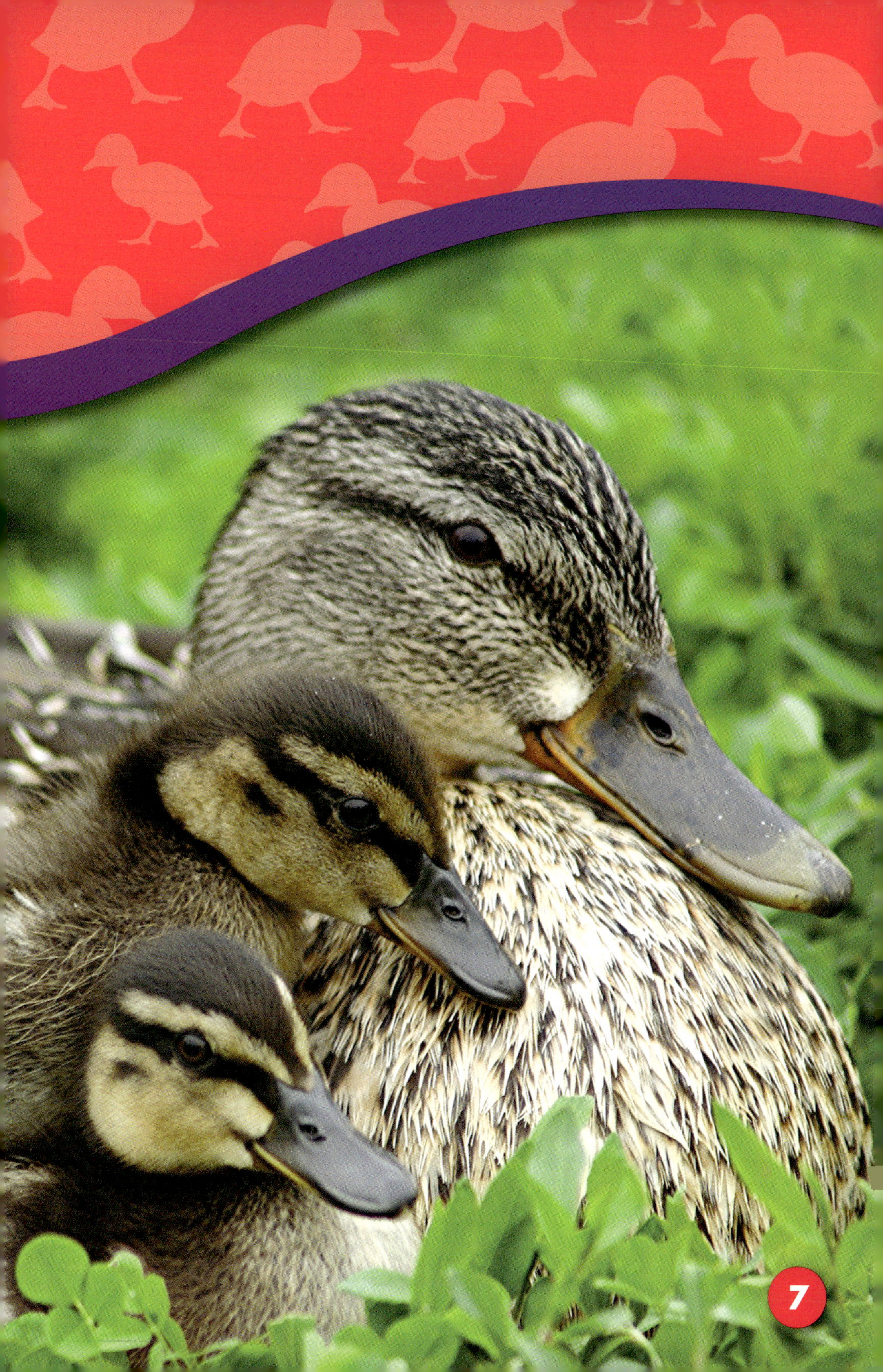

Ducklings have fuzzy feathers called **down**. Down can be yellow or brown.

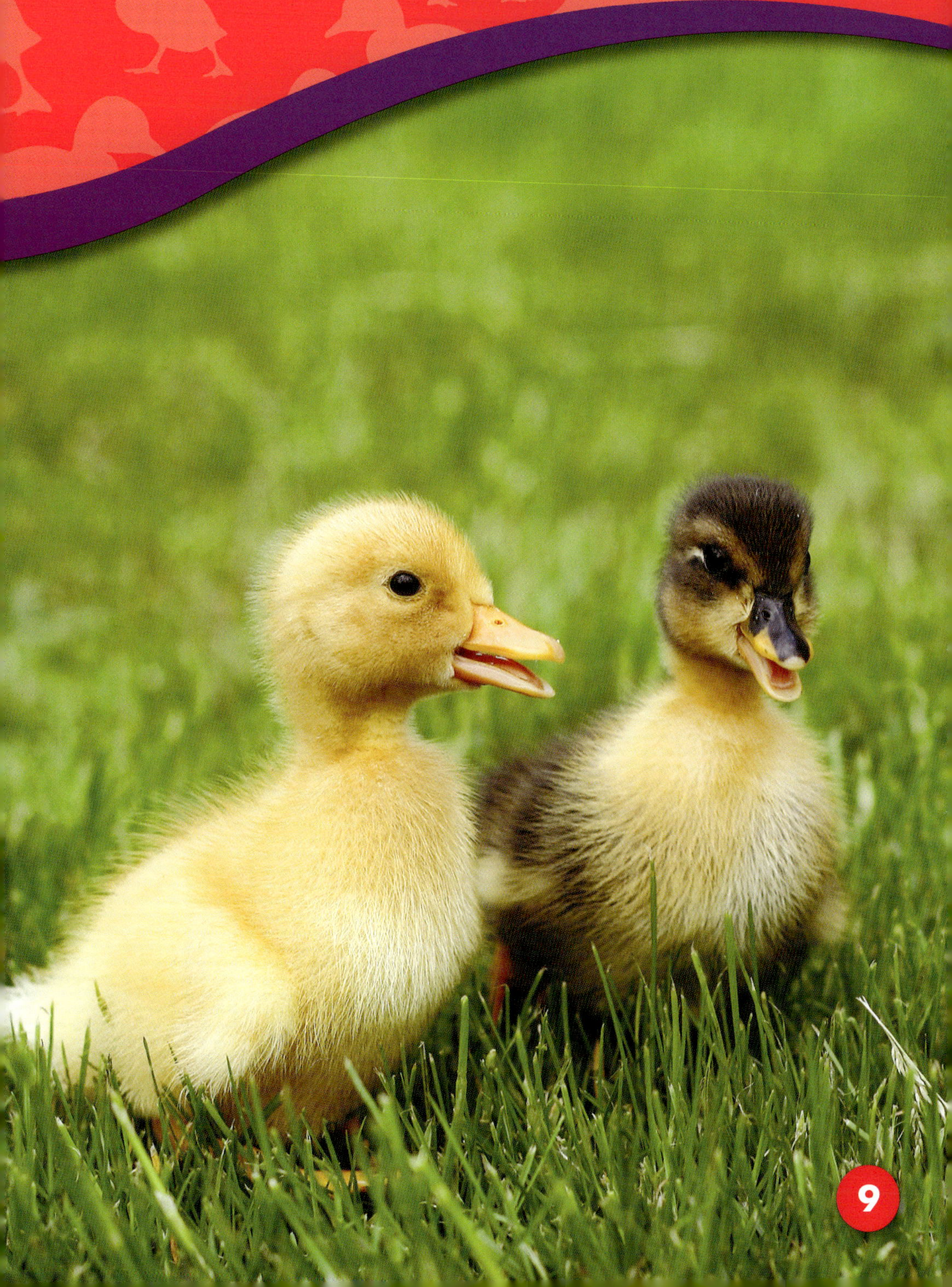

Ducklings have two small wings. The wings are not big enough to let a duckling fly.

11

Ducklings have two legs and **webbed feet**. Ducklings **paddle** their webbed feet to swim.

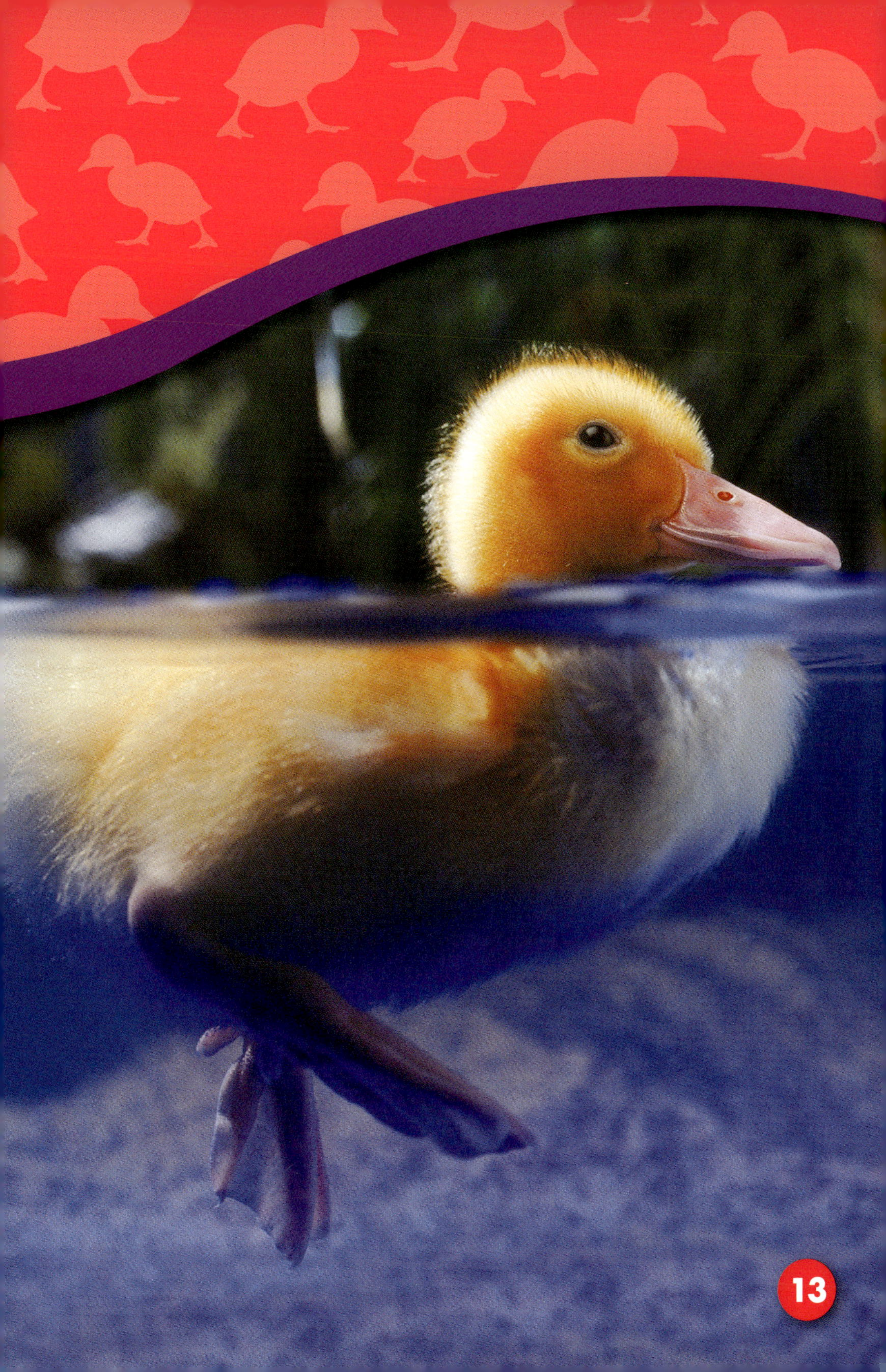

A duckling has a flat **bill**. This helps a duckling scoop up food.

bill

Ducklings find their own food. They eat seeds, **insects**, and small fish.

17

Ducklings stay near water as they grow.

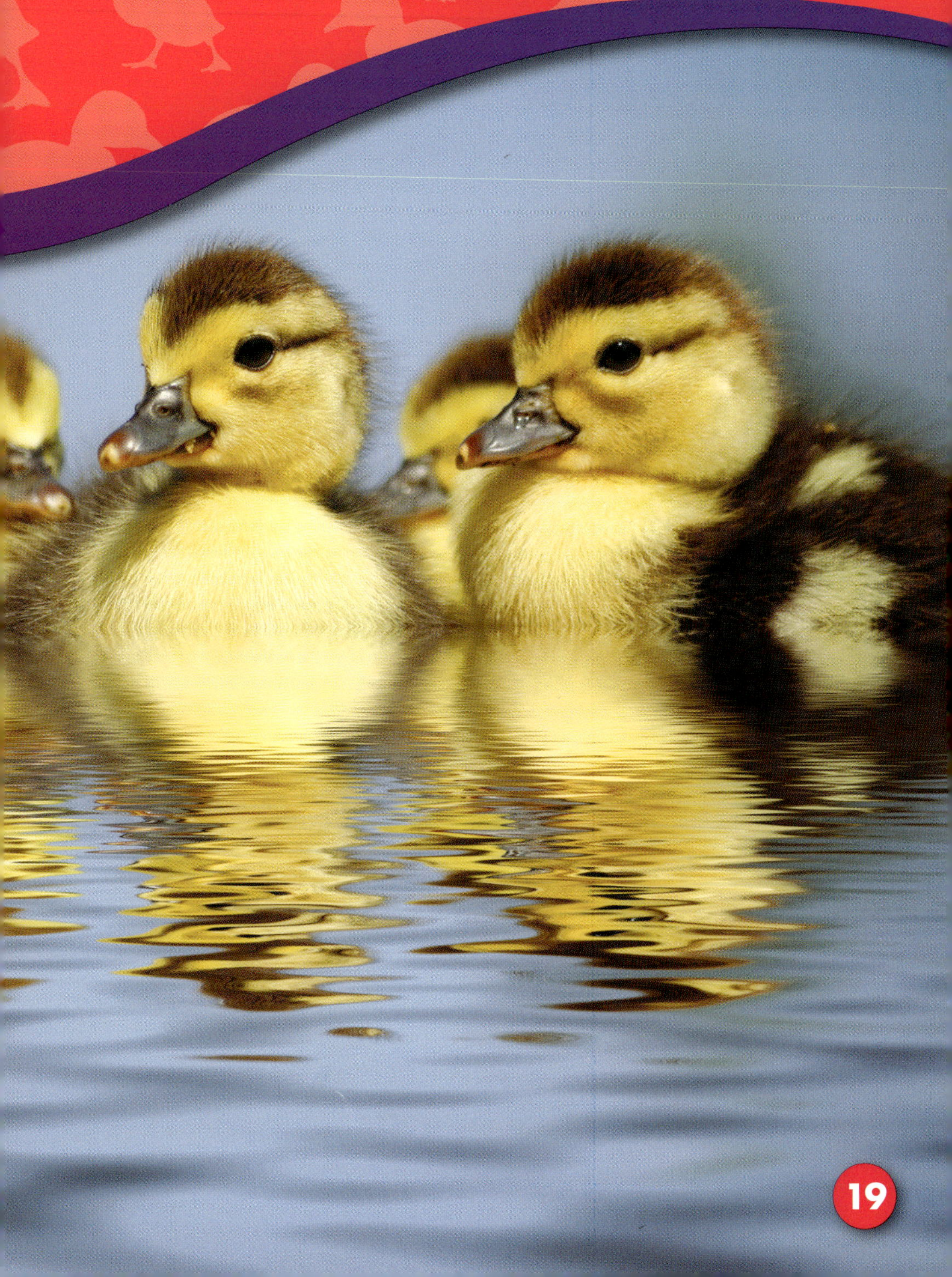

Ducklings grow fast.
Soon they have
adult feathers.
One day they may
have ducklings
of their own!

Glossary

bill—the hard mouth of a bird

down—the soft, fluffy feathers of newborn ducklings and other birds

hatch—to break out of an egg

insect—a small animal with six legs and a hard body divided into three parts

paddle—to move through water using arms or legs; ducklings use their feet to paddle.

webbed feet—feet with thin skin between the front toes; ducklings have three front toes and one back toe.

To Learn More

AT THE LIBRARY
Buzzeo, Toni. *Dawdle Duckling*. New York: Dial Books, 2003.

Hall, Margaret. *Ducks and Their Ducklings*. Mankato, Minn.: Pebble Books, 2004.

Jaffe, Elizabeth Dana. *Ducks Have Ducklings*. Minneapolis, Minn.: Compass Point Books, 2002.

Petty, Kate. *Ducklings*. Mankato, Minn.: Stargazer Books, 2006.

ON THE WEB
Learning more about ducklings is as easy as 1, 2, 3.

1. Go to www.factsurfer.com

2. Enter "ducklings" into search box.

3. Click the "Surf" button and you will see a list of related web sites.

With factsurfer.com, finding more information is just a click away.

Index

bill, 14
day, 20
down, 8
eggs, 4
feathers, 8, 20
fish, 16
food, 14, 16
insects, 16
legs, 12
mother, 6
seeds, 16
water, 18
webbed feet, 12
wings, 10

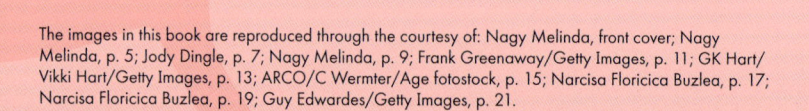

The images in this book are reproduced through the courtesy of: Nagy Melinda, front cover; Nagy Melinda, p. 5; Jody Dingle, p. 7; Nagy Melinda, p. 9; Frank Greenaway/Getty Images, p. 11; GK Hart/Vikki Hart/Getty Images, p. 13; ARCO/C Wermter/Age fotostock, p. 15; Narcisa Floricica Buzlea, p. 17; Narcisa Floricica Buzlea, p. 19; Guy Edwardes/Getty Images, p. 21.